KU-037-927

CONTENTS

EARLY BIRD
ENERGY

SOUND

BY SALLY M. WALKER
PHOTOGRAPHS BY ANDY KING

LER   Schools Library and Infomation Services   EAPOLIS

S00000689815

DUDLEY PUBLIC LIBRARIES

L

689815   SCH

J 534

Additional photographs are reproduced with permission from: © Howard Ande, p. 7; © Gerry Ellis, p.38; © Joe McDonald/CORBIS, p. 40; © Merlin D. Tuttle, Bat Conservation International, p. 41; © Lawrence Migdale/Photo Researchers, Inc., p. 42.

Text copyright © 2006 by Sally M. Walker
Photographs copyright © 2006 by Andy King

All rights reserved. International copyright secured. No part of this book may be reproduced, stored in a retrieval system, or transmitted in any form or by any means—electronic, mechanical, photocopying, recording, or otherwise—without the prior written permission of Lerner Publications Company, except for the inclusion of brief quotations in an acknowledged review.

Lerner Publications Company
A division of Lerner Publishing Group
241 First Avenue North
Minneapolis, MN 55401 U.S.A.

Website address: www.lernerbooks.com

Library of Congress Cataloging-in-Publication Data

Walker, Sally M.
    Sound / by Sally M. Walker ; photographs by Andy King.
       p.   cm. — (Early bird energy)
    Includes index.
    ISBN-13: 978–0–8225–2634–6 (lib. bdg. : alk. paper)
    ISBN-10: 0–8225–2634–4 (lib. bdg. : alk. paper)
    1. Sound—Juvenile literature. I. King, Andy, ill. II. Title. III. Series: Walker, Sally M. Early bird energy.
    QC225.5.W314 2006
    534—dc22   2005000028

Manufactured in the United States of America
1 2 3 4 5 6 – BP – 11 10 09 08 07 06

BE A WORD DETECTIVE

Can you find these words as you read about sound?
Be a detective and try to figure out what they mean.
You can turn to the glossary on page 46 for help.

atoms

molecule

sound waves

echo

noise

ultrasounds

infrasounds

pitch

vibrations

matter

reflected

We like some sounds better than other sounds. What do we call sounds that bother us?

CHAPTER 1
WHAT IS SOUND?

Clap your hands. Whisper your name. What do you hear? You hear a sound. Some sounds are nice to hear. But a sound like a fingernail scraping a chalkboard bothers many people. We call sounds that bother us noise.

Animals make sounds. So do machines. The wind blowing through trees makes a sound. What causes sound?

Sound begins when an object moves back and forth very quickly. The movements are called vibrations.

This excavator is part of a noisy construction site.

Loop a rubber band around your thumbs. Pluck the rubber band with your pinkie finger. This makes a sound.

Watch the band. You can see it vibrate. The vibrations move through the air to your ears. Then you hear the sound.

You hear a sound when you pluck a rubber band with your finger.

You can feel vibrations
when you hum.

You can feel vibrations too. Put one hand on top of your head and hum loudly. The tingling feeling in your hand is from the vibrations you make when you hum.

Sound goes from place to place by moving through matter. Anything that can be weighed or takes up space is called matter. Matter makes up everything around you. Matter can be a solid, like a table or a chair. Matter can be a liquid, like water. Or it can be a gas. The air you breathe is a gas.

Matter is all around you. This book is a solid.

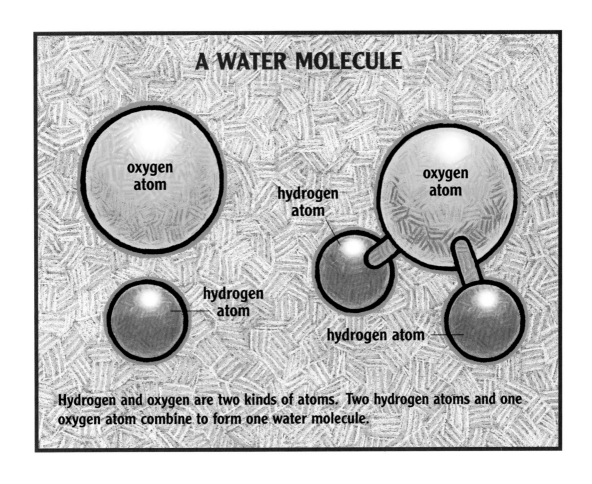

A WATER MOLECULE

oxygen atom

oxygen atom

hydrogen atom

hydrogen atom

hydrogen atom

Hydrogen and oxygen are two kinds of atoms. Two hydrogen atoms and one oxygen atom combine to form one water molecule.

Matter is made of tiny particles called atoms. A single atom is too small to see with your eyes. Billions of them can fit on the period at the end of this sentence. Atoms join together to make molecules. For example, one molecule of water has two atoms of hydrogen and one atom of oxygen.

Molecules are always moving. But molecules move in different ways in solids, liquids, and gases.

Molecules that are packed tightly together can't move very much. They form a solid. Tightly packed water molecules make ice. Ice is a solid.

These kids are very close together. They are like molecules in a solid.

These kids have lots of space to move around. They are like molecules in a gas.

Molecules that are less tightly packed make a liquid. Molecules in a liquid can move freely. That's why you can pour water from a glass into a bowl.

Molecules in a gas are spread out and move even more freely. They are so spread out that we can't see them. We can't see air. But we can see what happens when the air moves. Moving air is wind.

All matter is made of molecules. What do molecules do when something makes a sound?

CHAPTER 2
SOUND WAVES

The molecules in all kinds of matter vibrate when something makes a sound. The vibrations are called sound waves. A toy called a Slinky can show you how sound waves move.

Place the Slinky on a table. Stretch the ends about 2 feet apart. Quickly slide one hand a few inches toward the other hand.

The pictures on these pages show how a wave moves along a Slinky.

A wave will move along the Slinky. The wave pushes through each coil and makes it vibrate. The coil stops moving after the wave passes. Sound waves move from one molecule to the next just like the wave moves from one coil to the next.

A Slinky's wave moves in only one direction. But sound waves spread in all directions. They get weaker when they spread. Prove it yourself. Find a friend and stand about 25 feet apart. Whisper to your friend.

Your whisper's sound waves spread in all directions. Only some of the sound waves move toward your friend. Spreading makes them weak. That's why your friend can't hear you.

Sound waves spread out like the waves you see in a pool of water.

Do not shout through the hose, or you could hurt your friend's ear.

You can stop the sound waves from spreading out. You will need a garden hose. Make sure no water is inside the hose. Have your friend hold one end near his ear. You hold the other end. Whisper into the hose. Can your friend hear you?

The hose keeps your sound waves from spreading and becoming weaker. Almost all of them reach your friend's ear.

You can also make sound waves bounce. A ball thrown against a wall will bounce off. A sound wave will bounce off a hard surface too.

This ball bounces off a solid wall. So do sound waves.

You can make sound echo in an empty room.

Stand in the middle of an empty room. Shout. You will hear an echo of your shout. An echo is a reflected sound wave.

The sound waves from your shout moved from your mouth to the room's walls. Then the sound waves bounced off the walls and traveled back to you.

Some objects carry sound better than others. Find a metal pan, a kitchen towel, and a metal spoon.

Hold the handle of the metal pan. Tap the bottom of the pan with the spoon. It rings like a bell. The pan's metal molecules are tightly packed. Sound waves move strongly from one molecule to the next.

The metal pan and the metal spoon are both solids.

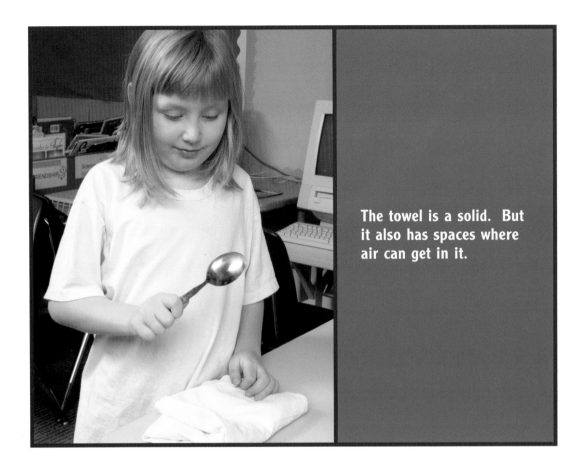

The towel is a solid. But it also has spaces where air can get in it.

Fold the towel and set it on a table. Tap it with the spoon. This time, you hear a thud. The molecules that make the towel are not as tightly packed as metal molecules. The molecules in the towel have air around them. The sound waves spread and weaken when they move through the towel.

You see lightning before you hear thunder because light waves travel faster than sound waves. How fast do sound waves travel?

CHAPTER 3
THE SPEED OF SOUND

How fast do sound waves move? A sound made 1 mile away from you takes about five seconds to reach you. It moves at about 1,096 feet per second. That's pretty fast!

Hot air molecules move quickly. They run into each other a lot, just like these kids.

Sound waves move faster on a hot day than they do on a cold day. Air molecules bump into each other more often when the air is hot. So sound waves can move from one molecule to the next more quickly.

Sound waves move at different speeds through liquids, solids, and gases. Sound waves travel 4 times faster in water than they do in air. That's because water molecules are closer together than air molecules. A sound wave can move from one water molecule to the next very quickly.

Sound waves move through water very quickly.

Sound waves move through wood even faster than they move through water.

Molecules of wood are even closer together than water molecules. So it takes even less time for a sound wave to move from one wood molecule to the next. Sound waves travel about 13 times faster in wood than they do in air.

Test the speed of sound in a solid wall. Ask a friend to stand at one end of a room. You stand at the other end. Ask your friend to tap the wall with a pen.

The sound you hear is not very loud. The air molecules between you and your friend carry the sound wave slowly. The sound has time to spread out and get weaker.

You do not need to tap the pen very hard for this experiment to work.

You can prove that sound moves faster in a solid than in a gas.

Put your ear against the wall. Listen again. The taps are much louder. Why? The sound waves move quickly through the wall's tightly packed molecules. The sound does not have time to spread out and get weak.

You can make a high sound with a rubber band. What makes the sound high or low?

CHAPTER 4
MAKING MUSIC

How fast an object vibrates changes the sound it makes. Usually, objects that vibrate fast make high sounds. We describe the high sound by saying it has a high pitch.

Stretch a rubber band tightly between your thumbs. Pluck it with your pinkie. The twang has a high pitch. The band vibrates very fast.

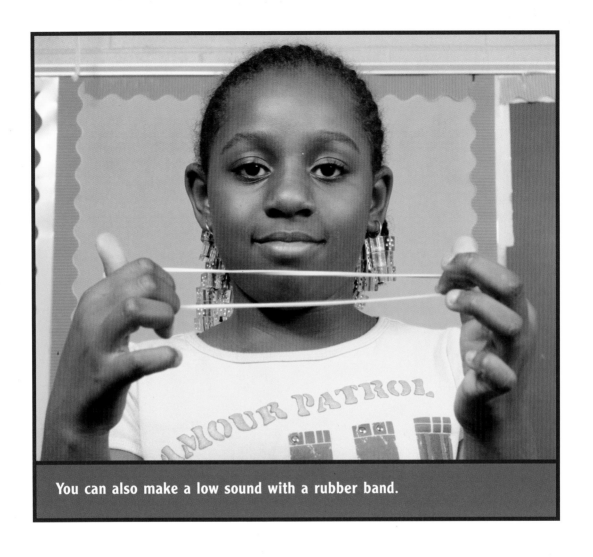
You can also make a low sound with a rubber band.

Move your thumbs together until the rubber band is hardly stretched. Pluck it again. The sound is much lower. That's because the band is vibrating more slowly. We describe the lower sound by saying it has a low pitch.

Music has sounds with many different pitches. You can blow across a drinking straw to make a sound. A short straw has a different pitch than a long one. Short straws have higher pitches. Long ones have lower pitches.

Blowing across a straw makes a sound. So does blowing across the opening of a bottle.

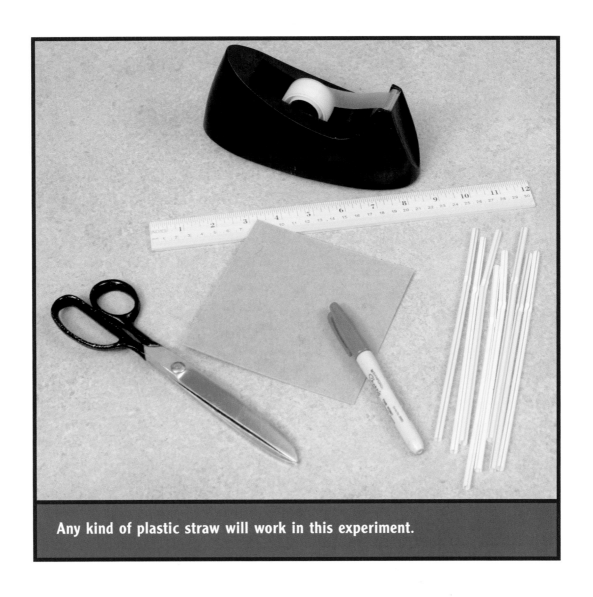

Any kind of plastic straw will work in this experiment.

Make a straw flute and prove it for
yourself. You will need eight plastic drinking
straws, a ruler, a marker, scissors, tape, and a
15-centimeter square of cardboard.

Measure one straw with the ruler. It is probably 19½ centimeters long. You can cut the straw if it is longer. Use your marker to make a line on the straw at 19½ centimeters. Cut the straw on the line. Write the number 1 on the straw.

Measure carefully so that your straw flute will sound good when you finish it.

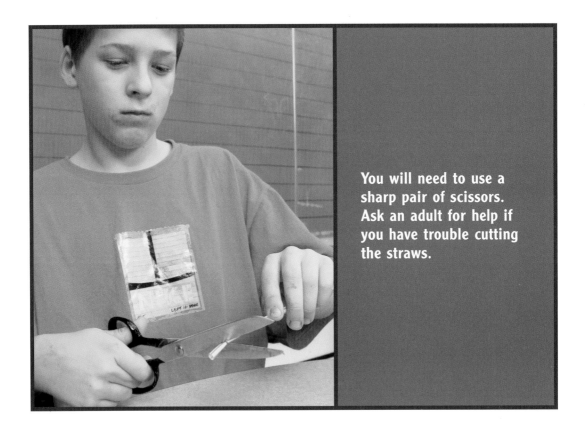

You will need to use a sharp pair of scissors. Ask an adult for help if you have trouble cutting the straws.

Measure, cut, and number the other seven straws. Their numbers and lengths should be as follows:

#2 = 18 centimeters

#3 = 16 centimeters

#4 = 14½ centimeters

#5 = 12½ centimeters

#6 = 11½ centimeters

#7 = 10 centimeters

#8 = 9½ centimeters

Tape the straws to the cardboard. Tape straw #1 along one edge of the cardboard. Make sure the straw sticks up a little bit over the top end of the cardboard. Tape straw #8 at the other edge of the cardboard. It should also stick up a little bit.

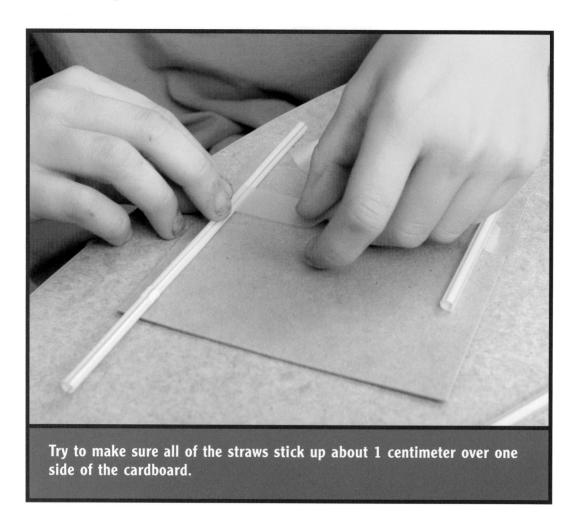

Try to make sure all of the straws stick up about 1 centimeter over one side of the cardboard.

Each straw will make a sound with a different pitch.

Put the rest of the straws on the cardboard between straws #1 and #8. The straws should be in order from longest to shortest. Tape the straws so that they stick up over the cardboard the same amount as straws #1 and #8.

Your flute is finished. Hold the cardboard with the ends of the straws near your bottom lip. Blow across the hole at the top of each straw. The pitch of each straw is different. Why? The air in each straw vibrates at a different speed when you blow. Try to play a song on your new flute.

When you blow across a straw, you make the air inside the straw vibrate. This makes a sound.

This girl can't hear her hands flapping. Why can't she hear them?

CHAPTER 5
SOUNDS WE CAN'T HEAR

People hear sound when an object vibrates. But there are some sounds we can't hear. We don't hear anything if an object vibrates less than 20 times per second.

You can wave your hand back and forth rapidly—but not faster than 20 times per second. So you don't hear a sound. The lowest notes on a piano are close to 20 vibrations per second.

Sounds made by vibrations of less than 20 times per second are called infrasounds. Thunderstorms and earthquakes can make infrasounds. We know about them because special instruments record and measure them.

Elephants make infrasounds to communicate with other elephants that are very far away.

Each key on the piano plays a note with a different vibration.

We can hear sounds made by objects that vibrate back and forth as many as 20,000 times per second. The very highest note on a piano vibrates about 15,000 times per second.

Sounds made by vibrations of more than 20,000 times per second are called ultrasounds. Bats make ultrasound waves.

Bats fly at night, when it is too dark to see. So they use ultrasound waves to help them. The ultrasounds hit objects such as the ground, tree branches, and insects. Then the sounds echo back to the bat.

Bats use sound to help them fly at night.

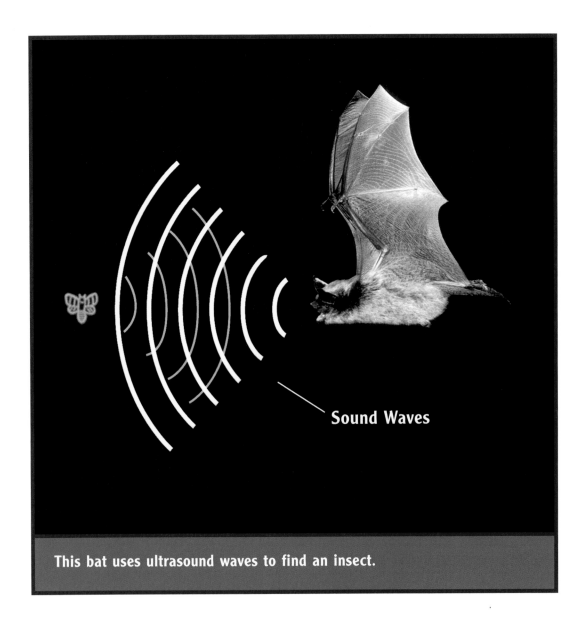

Sound Waves

This bat uses ultrasound waves to find an insect.

The echoes tell the bat where to fly. An ultrasound wave bounces off an insect. Then the bat can catch the insect and eat it!

You have learned a lot about sound.
Matter vibrates when a sound is made. Sound
travels in waves. Sounds can have high or
low pitches.

You can make the strings on a guitar vibrate. This makes music.

You hear many sounds all day long. What is your favorite sound?

Sit or stand still quietly for a minute. Close your eyes. Listen carefully. What can you hear? Hopefully, something that sounds good!

A NOTE TO ADULTS
ON SHARING A BOOK

When you share a book with a child, you show that reading is important. To get the most out of the experience, read in a comfortable, quiet place. Turn off the television and limit other distractions, such as telephone calls. Be prepared to start slowly. Take turns reading parts of this book. Stop occasionally and discuss what you're reading. Talk about the photographs. If the child begins to lose interest, stop reading. When you pick up the book again, revisit the parts you have already read.

BE A VOCABULARY DETECTIVE

The word list on page 5 contains words that are important in understanding the topic of this book. Be word detectives and search for the words as you read the book together. Talk about what the words mean and how they are used in the sentence. Do any of these words have more than one meaning? You will find the words defined in a glossary on page 46.

WHAT ABOUT QUESTIONS?

Use questions to make sure the child understands the information in this book. Here are some suggestions:

> What did this paragraph tell us? What does this picture show? What do you think we'll learn about next? What are vibrations? Does sound move faster through a gas or through a solid? Can you name some animals that hear sounds people don't hear? What is your favorite part of the book? Why?

If the child has questions, don't hesitate to respond with questions of your own, such as: What do *you* think? Why? What is it that you don't know? If the child can't remember certain facts, turn to the index.

INTRODUCING THE INDEX

The index helps readers find information without searching through the whole book. Turn to the index on page 48. Choose an entry such as *music,* and ask the child to use the index to find out how sound can be used to make music. Repeat with as many entries as you like. Ask the child to point out the differences between an index and a glossary. (The index helps readers find information, while the glossary tells readers what words mean.)

44

LEARN MORE ABOUT SOUND

BOOKS

Baker, Wendy, and Andrew Haslam. *Sound.* New York: Thompson Learning, 1995. Create complex sound experiments and musical instruments by following the instructions in this book.

Bradley, Kimberly Brubaker. *Energy Makes Things Happen.* New York: HarperCollins, 2003. This book describes the many kinds of energy all around us.

Henbest, Nigel, and Heather Couper. *Physics.* New York: Franklin Watts, 1983. This book introduces the basics of physics, the science of matter and energy.

Tocci, Salvatore. *Experiments with Sound.* New York: Children's Press, 2001. Learn about sound and the body, including how your ears and vocal chords work.

Wong, Ovid K. *Is Science Magic?* Chicago: Children's Press, 1989. This book is filled with experiments based on principles of matter and energy.

Wood, Robert W. *Sound Fundamentals: Funtastic Science Activities for Kids.* Philadelphia: Chelsea House Publishers, 1997. Try out 29 fun experiments and learn more about the science of sound.

WEBSITES

Neuroscience for Kids—Hearing Experiments
http://faculty.washington.edu/chudler/chhearing.html
This page includes information about the ear and hearing, along with fun sound experiments to try.

The Science of Music
http://www.exploratorium.edu/music/index.html
Learn more about music while using this fun site to make music of your own.

GLOSSARY

atoms: very tiny particles that come together and make molecules

echo: a sound that is heard again after sound waves hit a surface and bounce back

infrasounds: sounds that are too low-pitched for people to hear

matter: anything that takes up space and can be weighed. All things are made of matter.

molecule: the smallest amount of a substance that can be found

noise: a loud or harsh sound

pitch: how high or low a musical note sounds

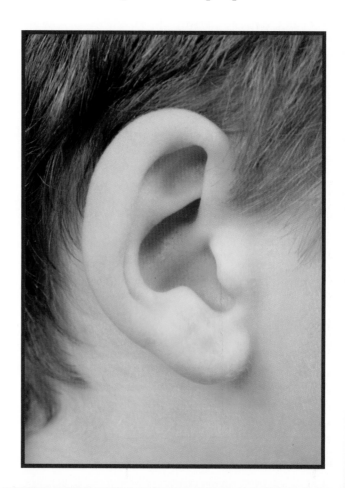

reflected: bounced back from a surface

sound waves: vibrations that can be heard or measured

ultrasounds: sounds that are too high-pitched for people to hear

vibrations: quick movements back and forth

INDEX

Pages listed in **bold** type refer to
photographs.

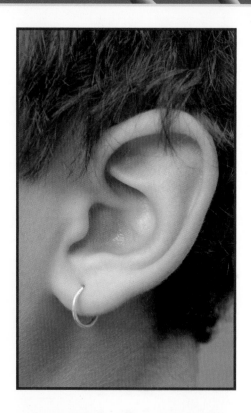

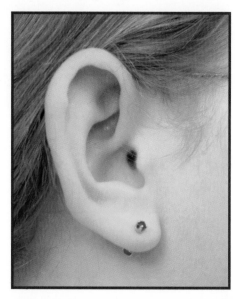